LINEAR IC APPLICATIONS

EDUCATIVE

I0494355

DR.C.ARUNABALA

XpressPublishing

An imprint of Notion Press

XpressPublishing
An imprint of Notion Press

Old No. 38, New No. 6
McNichols Road, Chetpet
Chennai - 600 031

First Published by Notion Press 2019
Copyright © Dr.C.Arunabala 2019
All Rights Reserved.

ISBN 978-1-64733-058-3

IC FABRICATION & CIRCUIT CONFIGURATION FOR LINEAR ICs

2 marks questions

1.Mention the advantages of integrated circuits.

- Miniaturization and hence increased equipment density.
- Cost reduction due to batch processing.
- Increased system reliability due to the elimination of soldered joints.
- Improved functional performance.
- Matched devices.
- Increased operating speeds.
- Reduction in power consumption.

2. Write down the various processes used to fabricate IC's using silicon planar technology.

- Silicon wafer preparation.
- Epitaxial growth
- Oxidation.
- Photolithography.
- Diffusion.
- Ion implantation.
- Isolation.
- Metallization.
- Assembly processing and packaging.

3. What is the purpose of oxidation?

- SiO_2 is an extremely hard protective coating and is unaffected by almost all reagents.
- By selective etching of SiO_2, diffusion of impurities through carefully defined windows can be accomplished to fabricate various components.

4. Why aluminium is preferred for metallization?

- It is a good conductor.

- It is easy to deposit aluminium films using vacuum deposition.
  - It makes good mechanical bonds with silicon.
    - It forms a low resistance contact.

5. What are the popular IC packages available?

- Metal can package.
- Dual-in-line package.
- Ceramic flat package.

6. Define an operational amplifier.

An operational amplifier is a direct-coupled, high gain amplifier consisting of one or more differential amplifier. By properly selecting the external components, it can be used to perform a variety of mathematical operations.

7. Mention the characteristics of an ideal op-amp.

- Open loop voltage gain is infinity.
- Input impedance is infinity.
- Output impedance is zero.
- Bandwidth is infinity.
- Zero offset.

8. What happens when the common terminal of V+ and V- sources is not grounded?

If the common point of the two supplies is not grounded, twice the supply voltage will get applied and it may damage the op-amp.

9. Define input offset voltage.

A small voltage applied to the input terminals to make the output voltage as zero when the two input terminals are grounded is called input offset voltage.

10. Define input offset current. State the reasons for the offset currents at the input of the op-amp.

The difference between the bias currents at the input terminals of the op-amp is called as input offset current. The input terminals conduct a small value of dc current to bias the input transistors. Since the input transistors cannot be made identical, there exists a difference in bias currents.

11. Define CMRR of an op-amp.

The relative sensitivity of an op-amp to a difference signal as compared to a common –mode signal is called the common –mode rejection ratio. It is expressed in decibels.

$$CMRR = Ad/Ac$$

12.What are the applications of current sources?

Transistor current sources are widely used in analog ICs both as biasing elements and as load devices for amplifier stages.

13. Justify the reasons for using current sources in integrated circuits.

- Superior insensitivity of circuit performance to power supply variations and temperature.
- More economical than resistors in terms of die area required to provide bias currents of small value.
- When used as load element, the high incremental resistance of current source, results in high voltage gain at low supply voltages.

14. What is the advantage of widlar current source over constant current source?

Using constant current source output current of small magnitude(microamp range) is not attainable due to the limitations in chip area. Widlar current source is useful for obtaining small output currents.Sensitivity of widlar current source is less compared to constant current source.

15.Mention the advantages of Wilson current source.

- Provides high output resistance.
- Offers low sensitivity to transistor base currents.

16. Define sensitivity.

Sensitivity is defined as the percentage or fractional change in output current per percentage or fractional change in power-supply voltage.

17. What are the limitations in a temperature compensated zener-reference source?

A power supply voltage of atleast 7 to 10 V is required to place the diode in the breakdown region and that substantial noise is introduced in the circuit by the avalanching diode.

18. What do you mean by a band-gap referenced biasing circuit?

The biasing sources referenced to VBE has a negative temperature co-efficient and VT has a positive temperature co-efficient. Band gap reference circuit is one in which the output current is referenced to a composite voltage that is a weighted sum of VBE and VT so that by proper weighting, zero temperature co efficient can be achieved.

19. In practical op-amps, what is the effect of high frequency on its performance?

The open-loop gain of op-amp decreases at higher frequencies due to the presence of parasitic capacitance. The closed-loop gain increases at higher frequencies and leads to instability.

20. What is the need for frequency compensation in practical op-amps?

Frequency compensation is needed when large bandwidth and lower closed loop gain is desired. Compensating networks are used to control the phase shift and hence to improve the stability.

21. Mention the frequency compensation methods.

Dominant-pole compensation and Pole-zero compensation.

22. What are the merits and demerits of Dominant-pole compensation?

- Noise immunity of the system is improved.
- Open-loop bandwidth is reduced.

23. Define slew rate.

The slew rate is defined as the maximum rate of change of output voltage caused by a step input voltage. An ideal slew rate is infinite which means that op-amp's output voltage should change instantaneously in response to input step voltage.

24. Why IC 741 is not used for high frequency applications?

IC741 has a low slew rate because of the predominance of capacitance present in the circuit at higher frequencies. As frequency increases the output gets distorted due to limited slew rate.

25. What causes slew rate?

There is a capacitor with-in or outside of an op-amp to prevent oscillation. It is this capacitor which prevents the output voltage from responding immediately to a fast changing input.

26. Obtain the frequency response of an open-loop op-amp and discuss about the methods of frequency compensation .

Ans:

The open-loop gain of op-amp decreases at higher frequencies due to the presence of parasitic capacitance. The closed-loop gain increases at higher frequencies and leads to instability. Frequency compensation is needed when large bandwidth and lower closed loop gain is desired. Compensating networks are used to control the phase shift and hence to improve the stability.

Frequency compensation methods:

- Dominant-pole compensation
- Pole-zero compensation.

APPLICATIONS OF OP – AMPS

2 Marks Questions:

1. Mention some of the linear applications of op – amps :

Adder, subtractor, voltage –to- current converter, current –to- voltage converters, instrumentation amplifier, analog computation ,power amplifier, etc are some of the linear op-amp circuits.

2. Mention some of the non – linear applications of op-amps:-

Rectifier, peak detector, clipper, clamper, sample and hold circuit, log amplifier, anti –log amplifier, multiplier are some of the non – linear op-amp circuits.

3. What are the areas of application of non-linear op- amp circuits?

- Industrial instrumentation
- Communication
- Signal processing

4. What is the need for an instrumentation amplifier?

In a number of industrial and consumer applications, the measurement of physical quantities is usually done with the help of transducers. The output of transducer has to be amplified So that it can drive the indicator or display system. This function is performed by an instrumentation amplifier.

5. List the features of instrumentation amplifier:

- high gain accuracy
- high CMRR
- high gain stability with low temperature co-efficient
- low dc offset
- low output impedance

6. What are the applications of V-I converter?

- Low voltage dc and ac voltmeter

- L E D
  - Zener diode tester

7. What do you mean by a precision diode?

The major limitation of ordinary diode is that it cannot rectify voltages below the cut – in voltage of the diode. A circuit designed by placing a diode in the feedback loop of an op – amp is called the precision diode and it is capable of rectifying input signals of the order of millivolt.

8. Write down the applications of precision diode.

- Half - wave rectifier
- Full - Wave rectifier
- Peak – value detector
- Clipper
- Clamper

9. List the applications of Log amplifiers:

- Analog computation may require functions such as lnx, log x, sin hx etc. These functions can be performed by log amplifiers
- Log amplifier can perform direct dB display on digital voltmeter and spectrum analyzer
- Log amplifier can be used to compress the dynamic range of a signal

10. What are the limitations of the basic differentiator circuit?

- At high frequency, a differentiator may become unstable and break into oscillations
- The input impedance decreases with increase in frequency, thereby making the circuit sensitive to high frequency noise.

11. Write down the condition for good differentiation :-

For good differentiation, the time period of the input signal must be greater than or equal to $R_f C_1$

$$T > Rf \, C1$$

Where, Rf is the feedback resistance

Cf is the input capacitance

12. What is a comparator?

A comparator is a circuit which compares a signal voltage applied at one input of an op-amp with a known reference voltage at the other input. It is an open loop op - amp with output + Vsat .

13. What are the applications of comparator?

- Zero crossing detector
- Window detector
- Time marker generator
- Phase detector

14. What is a Schmitt trigger?

Schmitt trigger is a regenerative comparator. It converts sinusoidal input into a square wave output. The output of Schmitt trigger swings between upper and lower threshold voltages, which are the reference voltages of the input waveform.

15. What is a multivibrator?

Multivibrators are a group of regenerative circuits that are usedextensively in timing applications. It is a wave shaping circuit which gives symmetric orasymmetric square output. It has two states either stable or quasi- stable depending onthe type of multivibrator.

16. What do you mean by monostable multivibrator?

Monostable multivibrator is one which generates a single pulse of specified duration in response to each external trigger signal. It has only one stable state. Application of a trigger causes a change to the quasi-stable state. An external trigger signal generated due to charging and discharging of the capacitor produces the transition to the original stable state.

17. What is an astable multivibrator?

Astable multivibrator is a free running oscillator having two quasi-stable states. Thus, there are oscillations between these two states and no external signal is required to produce the change in state.

18. What is a bistable multivibrator?

Bistable multivibrator is one that maintains a given output voltage level unless an external trigger is applied. Application of an external trigger signal causes a change of state, and this output level is maintained indefinitely until a second trigger is applied. Thus, it requires two external triggers before it returns to its initial state.

19. What are the requirements for producing sustained oscillations in feedback circuits?

For sustained oscillations,

- The total phase shift around the loop must be zero at the desired frequency of oscillation, fo. ie, $\angle AB = 0$ (or) 360°
- At fo, the magnitude of the loop gain $|A b|$ should be equal to unity

20. Mention any two audio frequency oscillators:

- RC phase shift oscillator
- Wein bridge oscillator

21. What are the characteristics of a comparator?

- Speed of operation
- Accuracy
- Compatibility of the output

23. What are the demerits of passive filters?

Passive filters works well for high frequencies. But at audio frequencies, the inductors become problematic, as they become large, heavy and expensive. For low frequency applications, more number of turns of wire must be used which in turn adds to the series resistance degrading inductor's performance ie, low Q, resulting in high power dissipation.

24. What are the advantages of active filters?

Active filters used op- amp as the active element and resistors and capacitors as passive elements.

- By enclosing a capacitor in the feedback loop , inductor less active fulters can be obtained.
- Op-amp used in non – inverting configuration offers high input impedance and low output impedance, thus improving the load drive capacity.

25. Mention some commonly used active filters:

- Low pass filter
- High pass filter
- Band pass filter
- Band reject filter

ANALOG MULTIPLIERS AND PLL

2 marks questions

1. Mention some areas where PLL is widely used:

- Radar synchronization
- satellite communication systems
- air borne navigational systems
- FM communication systems
- Computers.

2. List the basic building blocks of PLL:

- Phase detector/comparator
  - Low pass filter
  - Error amplifier
- Voltage controlled oscillator

3. What are the three stages through which PLL operates?

- Free running
  - Capture
- Locked/ tracking

4. Define lock-in range of a PLL:

The range of frequencies over which the PLL can maintain lock with the incoming signal is called the lock-in range or tracking range. It is expressed as a percentage of the VCO free running frequency.

5. Define capture range of PLL:

The range of frequencies over which the PLL can acquire lock with an input signal is called the capture range. It is expressed as a percentage of the VCO free running frequency.

6. Define Pull-in time.

The total time taken by the PLL to establish lock is called pull-in time. It depends on the initial phase and frequency difference between the two signals as well as on the overall loop gain and loop filter characteristics.

7. For perfect lock, what should be the phase relation between the incoming signal and VCO output signal?

The VCO output should be 90 degrees out of phase with respect to the input signal.

8. Give the classification of phase detector

- Analog phase detector
- Digital phase detector

9. What is a switch type phase detector?

An electronic switch is opened and closed by signal coming from VCO and the input signal is chopped at a repetition rate determined by the VCO frequency. This type of phase detector is called a half wave detector since the phase information for only one half of the input signal is detected and averaged.

10. What are the problems associated with switch type phase detector?

The output voltage Ve is proportional to the input signal amplitude. This is undesirable because it makes phase detector gain and loop gain dependent on the input signal amplitude.

The output is proportional to $\cos\varphi$ making it non linear.

11. What is a voltage controlled oscillator?

Voltage controlled oscillator is a free running multivibrator operating at a set frequency called the free running frequency. This frequency can be shifted to either side by applying a dc control voltage and the frequency deviation is proportional to the dc control voltage.

12. On what parameters does the free running frequency of VCO depend on?

- External timing resistor, RT
- External timing capacitor, CT
- The dc control voltage Vc.

13. Give the expression for the VCO free running frequency.

$$f_o = 0.25 / R_T C_T$$

14. Define Voltage to Frequency conversion factor.

Voltage to Frequency conversion factor is defined as,

$$K_v = \Delta f_o/\Delta V_c = V_{cc}$$

Where, ΔV_c is the modulation voltage required to produce the frequency shift Δfo

15. What is the purpose of having a low pass filter in PLL?

- It removes the high frequency components and noise.
- Controls the dynamic characteristics of the PLL such as capture range, lock-in range, band-width and transient response.

The charge on the filter capacitor gives a short- time memory to the PLL

16. Discuss the effect of having large capture range.

The PLL cannot acquire a signal outside the capture range, but once captured, it will hold on till the frequency goes beyond the lock-in range. Thus , to increase the ability of lock range, large capture range is required. But, a large capture range will make the PLL more susceptible to noise and undesirable signal.

17. Mention some typical applications of PLL:

· Frequency multiplication/division

· Frequency translation

· AM detection

· FM demodulation

· FSK demodulation.

18. What is a compander IC? Give some examples.

The term companding means compressing and expanding. In a communication system, the audio signal is compressed in the transmitter and expanded in the receiver. Examples: LM 2704- LM 2707 ; NE 570/571.

19. What are the merits of companding?

- The compression process reduces the dynamic range of the signal before it is transmitted.
- Companding preserves the signal to noise ratio of the original signal and avoids non linear distortion of the signal when the input amplitude is large.
- It also reduces buzz, bias and low level audio tones caused by mild interference.

20. List the applications of OTA:

OTA can be used in

· programmable gain voltage amplifier

· sample and hold circuits

· voltage controlled state variable filter

· current controlled relaxation oscillator.

21.Applications of PLL:

AM detection

FM demodulation

FSK demodulation

Frequency multiplication/division.

A/D AND D/A CONVERTERS

2 marks questions

1. List the broad classification of ADCs.

1. Direct type ADC.

2. Integrating type ADC.

2. List out the direct type ADCs.

1. Flash (comparator) type converter

2. Counter type converter

3. Tracking or servo converter

4. Successive approximation type converter

3. List out some integrating type converters.

1. Charge balancing ADC

2. Dual slope ADC

4. What is integrating type converter?

An ADC converter that perform conversion in an indirect manner by first changing the analog I/P signal to a linear function of time or frequency and then to a digital code is known as integrating type A/D converter.

5. Explain in brief the principle of operation of successive Approximation ADC.

The circuit of successive approximation ADC consists of a successive approximation register (SAR), to find the required value of each bit by trial & error. With the arrival of START command, SAR sets the MSB bit to 1. The O/P is converted into an analog signal & it is compared with I/P signal. This O/P is low or High. This process continues until all bits are checked.

6. What are the main advantages of integrating type ADCs?

i. The integrating type of ADC's doing not need a sample/Hold circuit at the input.

ii. It is possible to transmit frequency even in noisy environment or in an isolated form.

7. Where are the successive approximation type ADC's used?

The Successive approximation ADCs are used in applications such as data loggers & instrumentation where conversion speed is important.

8. What is the main drawback of a dual-slop ADC?

The dual slope ADC has long conversion time. This is the main drawback of dual slope ADC.

9. State the advantages of dual slope ADC:

It provides excellent noise rejection of ac signals whose periods are integral multiples of the integration time T.

10. Define conversion time.

It is defined as the total time required to convert an analog signal into its digital output. It depends on the conversion technique used & the propagation delay of circuit components. The conversion time of a successive approximation type ADC is given by

$$T(n+1)$$

where T---clock period

Tc---conversion time

n----no. of bits

11. Define resolution of a data converter.

The resolution of a converter is the smallest change in voltage which may be produced at the output or input of the converter.

Resolution (in volts)= VFS/2n-1=1 LSB increment. The resolution of an ADC is defined as the smallest change in analog input for a one bit change at the output.

12. Define accuracy of converter.

Absolute accuracy:

It is the maximum deviation between the actual converter output & the ideal converter output.

Relative accuracy:

It is the maximum deviation after gain & offset errors have been removed. The accuracy of a converter is also specified in form of LSB increments or % of full scale voltage.

13. What is settling time?

It represents the time it takes for the output to settle within a specified band $\pm\frac{1}{2}$LSB of its final value following a code change at the input (usually a full scale change). It depends upon the switching time of the logic circuitry due to internal parasitic capacitance & inductances. Settling time ranges from 100ns. 10µs depending on word length & type circuit used.

14. Explain in brief stability of a converter:

The performance of converter changes with temperature age & power supply variation. So all the relevant parameters such as offset, gain, linearity error & monotonicity must be specified over the full temperature & power supply ranges to have better stability performances.

15. What is meant by linearity?

The linearity of an ADC/DAC is an important measure of its accuracy & tells us how close the converter output is to its ideal transfer characteristics. The linearity error is usually expressed as a fraction of LSB increment or percentage of full-scale voltage. A good converter exhibits a linearity error of less than $\pm\frac{1}{2}$ LSB.

16. What is monotonic DAC?

A monotonic DAC is one whose analog output increases for an increase in digital input.

17. What is multiplying DAC?

A digital to analog converter which uses a varying reference voltage VR is called a multiplying DAC (MDAC). If the reference voltage of a DAC, VR is a sine wave give by

$$V(t) = VinCos2\pi ft$$

$$\text{Then, } Vo(t) = Vom\,Cos(2\pi ft + 180°)$$

18. What is a sample and hold circuit? Where it is used?

A sample and hold circuit is one which samples an input signal and holds on to its last sampled value until the input is sampled again. This circuit is mainly used in digital interfacing, analog to digital systems, and pulse code modulation systems.

19. Define sample period and hold period.

The time during which the voltage across the capacitor in sample and hold circuit is equal to the input voltage is called sample period. The time period during which the voltage across the capacitor is held constant is called hold period.

20. What is meant by delta modulation?

Delta modulation is a technique capable of performing analog signal quantisation with smaller bandwidth requirements. Here, the binary output representing the most recent sampled amplitude will be determined on the basis of previous sampled amplitude levels.

21. What is integrating type converter?

An ADC converter that perform conversion in an indirect manner by first changing the analog I/P signal to a linear function of time or frequency and then to a digital code is known as integrating type A/D converter.

22. Explain the principle of operation of successive Approximation ADC.

The circuit of successive approximation ADC consists of a successive approximation register (SAR), to find the required value of each bit by trial & error. With the arrival of START command, SAR sets the MSB bit to 1. The O/P is converted into an analog signal & it is compared with I/P signal. This O/P is low or High. This process continues until all bits are checked. Functional diagram Operation, Truth table, Output graph.

23. What are various types of digital to analog converters:

Weighted resistor DAC

R-2R ladder DAC

Inverted R-2R ladder DAC

Circuit diagram & operation for each

24. What is delta sigma modulation? Explain the A/D conversion using Delta modulator.

Delta modulation is a technique capable of performing analog signal quantisation with smaller bandwidth requirements. Here, the binary output representing the most recent sampled amplitude will be determined on the basis of previous sampled amplitude levels. Functional diagram Operation.

WAVEFORM GENERATORS & SPECIAL FUNCTION ICs

2 mark questions

1. Mention some applications of 555 timer:

- Oscillator
- pulse generator
- ramp and square wave generator
- mono-shot multivibrator
- burglar alarm
- traffic light control.

2. List the applications of 555 timer in monostable mode of operation:

- missing pulse detector
- Linear ramp generator
- Frequency divider
- Pulse width modulation.

3. List the applications of 555 timer in Astable mode of operation:

- FSK generator
- Pulse-position modulator

4. What is a voltage regulator?

A voltage regulator is an electronic circuit that provides a stable dc voltage independent of the load current, temperature, and ac line voltage variations.

5. Give the classification of voltage regulators:

- Series / Linear regulators
- Switching regulators.

6.What is a linear voltage regulator?

Series or linear regulator uses a power transistor connected in series between the unregulated dc input and the load and it conducts in the linear region .The output voltage is controlled by the continous voltage drop taking place across the series pass transistor.

7. What is a switching regulator?

Switching regulators are those which operate the power transistor as a high frequency on/off switch, so that the power transistor does not conduct current continously.This gives improved efficiency over series regulators.

8. What are the advantages of IC voltage regulators?

- low cost
- high reliability
- reduction in size
- excellent performance

9. Give some examples of monolithic IC voltage regulators:

- 78XX series fixed output, positive voltage regulators
- 79XX series fixed output, negative voltage regulators
- 723 general purpose regulator.

10.What is the purpose of having input and output capacitors in three terminal IC regulators?

A capacitor connected between the input terminal and ground cancels the inductive effects due to long distribution leads. The output capacitor improves the transient response.

11. Define line regulation.

Line regulation is defined as the percentage change in the output voltage for a change in the input voltage.It is expressed in millivolts or as a percentage of the output voltage.

12. Define load regulation.

Load regulation is defined as the change in output voltage for a change in load current. It is expressed in millivolts or as a percentage of the output voltage.

13. What is meant by current limiting?

Current limiting refers to the ability of a regulator to prevent the load current from increasing above a preset value.

14. Give the drawbacks of linear regulators:

The input step down transformer is bulky and expensive because of low line frequency.

Because of low line frequency,large values of filter capacitors are required to decrease the ripple.

Efficiency is reduced due to the continous power dissipation by the transistor as it operates in the linear region.

15. What is the advantage of switching regulators?

Greater efficiency is achieved as the power transistor is made to operate as low impedance switch.Power transmitted across the transistor is in discrete pulses rather than as a steady current flow.

By using suitable switching loss reduction technique, the switching frequency can be increased so as to reduce the size and weight of the inductors and capacitors.

16. What is an opto-coupler IC? Give examples.

Opto-coupler IC is a combined package of a photo-emitting device and a photosensing device.Examples for opto-coupler circuit : LED and a photo diode, LED and photo transistor, LED and Darlington.

Examples for opto-coupler IC : MCT 2F , MCT 2E .

17. Mention the advantages of opto-couplers:

- Better isolation between the two stages.
- Impedance problem between the stages is eliminated.
- Wide frequency response.
- Easily interfaced with digital circuit.
- Compact and light weight.
- Problems such as noise, transients, contact bounce are eliminated.

18. What is an isolation amplifier?

An isolation amplifier is an amplifier that offers electrical isolation between its input and output terminals.

19. What is the need for a tuned amplifier?

In radio or TV receivers , it is necessary to select a particular channel among all other available channels.Hence some sort of frequency selective circuit is needed that will allow us to amplify the frequency band required and reject all the other unwanted signals and this function is provided by a tuned amplifier.

20. Give the classification of tuned amplifier:

(i) Small signal tuned amplifier

- Single tuned

- Double tuned
- Stagger tuned

(ii) Large signal tuned amplifier.

IC Fabrication

1. Define an Integrated circuit.

An integrated circuit(IC) is a miniature, low cost electronic circuit consisting of active and passive components fabricated together on a single crystal of silicon. The active components are transistors and diodes and passive components are resistors and capacitors.

2. What are the basic processes involved in fabricating ICs using planar technology?

- Silicon wafer (substrate) preparation
- Epitaxial growth
- Oxidation
- Photolithography
- Diffusion
- Ion implantation
- Isolation technique
- Metallization
- Assembly processing & packaging

Characteristics of Op-Amp

1. What are the advantages of ICs over discrete circuits?

1. Minimization & hence increased equipment density.

2. Cost reduction due to batch processing.

3. Increased system reliability

4. Improved functional performance.

5. Matched devices.

6. Increased operating speeds

7. Reduction in power consumption

2. What is OPAMP?

An operational amplifier is a direct coupled high gain amplifier consisting of one or more differential amplifiers , followed by a level translator and an output stage .It is a versatile device that can be used to amplify ac as well as dc input signals & designed for computing mathematical functions such as addition, subtraction , multiplication, integration & differentiation.

3. List out the ideal characteristics of OPAMP?

- Open loop gain infinite
- Input impedance infinite
- Output impedance low
- Bandwidth infinite
- Zero offset, i.e., $V_o=0$ when $V_1=V_2=0$

4. What are the different kinds of packages of IC741?

a) Metal can (TO) package

b) Dual-in-line package

c) Flat package or flat pack

5. What are the assumptions made from ideal op – amp characteristics?

- The current drawn by either of the input terminals (noninverting/ inverting) is negligible.

- The potential difference between the inverting & non-inverting input terminals is zero.

6. Mention some of the linear applications of op – amps

Adder, subtractor, voltage –to- current converter, current –to- voltage converters, instrumentation amplifier, analog computation, power amplifier, etc are some of the linear op-amp circuits.

7. Mention some of the non – linear applications of op-amps

Rectifier, peak detector, clipper, clamper, sample and hold circuit, log amplifier, anti –log amplifier, multiplier are some of the non – linear op-amp circuits.

8. What are the areas of application of non-linear op- amp circuits?

‾ industrial instrumentation

‾ Communication

‾ Signal processing

9. What happens when the common terminal of V+ and V- sources is not grounded?

If the common point of the two supplies is not grounded, twice the supply voltage will get applied and it may damage the op-amp.

10. Define input offset voltage.

A small voltage applied to the input terminals to make the output voltage as zero when the two input terminals are grounded is called input offset voltage.

11. Define input offset current. State the reasons for the offset currents at the input of the op-amp.

The difference between the bias currents at the input terminals of the op-amp is called as input offset current. The input terminals conduct a small value of dc current to bias the input transistors. Since the input transistors cannot be made identical. There exists a difference in bias currents.

12. Define CMRR of an op-amp.

The relative sensitivity of an op-amp to a difference signal as compared to a common –mode signal is called the common – mode rejection ratio. It is expressed in decibels.

$$CMRR = Ad/Ac$$

13. In practical op-amps, what is the effect of high frequency on its performance?

The open-loop gain of op-amp decreases at higher frequencies due to the presence of parasitic capacitance. The closed-loop gain increases at higher frequencies and leads to instability.

14. What is the need for frequency compensation in practical op-amps?

Frequency compensation is needed when large bandwidth and lower closed loop gain is desired. Compensating networks are used to control the phase shift and hence to improve the stability.

15. Mention the frequency compensation methods.

* Dominant-pole compensation

* Pole-zero compensation.

16. What are the merits and demerits of Dominant-pole compensation?

- Noise immunity of the system is improved.
- Open-loop bandwidth is reduced.

17. Define slew rate.

The slew rate is defined as the maximum rate of change of output voltage caused by a step input voltage. An ideal slew rate is infinite which means that op-amp's output voltage should change instantaneously in response to input step voltage.

18. Why IC 741 is not used for high frequency applications?

IC741 has a low slew rate because of the predominance of capacitance present in the circuit at higher frequencies. As frequency increases the output gets distorted due to limited slew rate.

19. What causes slew rate

There is a capacitor with-in or outside of an op-amp to prevent oscillation. It is this capacitor which prevents the output voltage from responding immediately to a fast changing input.

20. Define thermal drift.

The bias current, offset current & offset voltage change with temperature. A circuit carefully nulled at 25oC may not remain so when the temperature raises to 35oC.This is called thermal drift . Often, offset current drift is expressed in nA/ °C and offset voltage drift in mV/ °C.

21. Define supply voltage rejection ratio (SVRR)

The change in OPAMP's input offset voltage due to variations in supply voltage is called the supply voltage rejection ratio. It is also called Power Supply Rejection Ratio (PSRR) or Power Supply Sensitivity (PSS).

Applications of Op Amp

1. What is the need for an instrumentation amplifier?

In a number of industrial and consumer applications, the measurement of physical quantities is usually done with the help of transducers. The output of transducer has to be amplified So that it can drive the indicator or display system. This function is performed by an instrumentation amplifier.

2. List the features of instrumentation amplifier

- high gain accuracy

- high CMRR

- high gain stability with low temperature co-efficient

- low dc offset

- low output impedance

3. What is a comparator?

A comparator is a circuit which compares a signal voltage applied at one input of an op-amp with a known reference voltage at the other input. It is an open loop op -amp with output +Vsat .

4. What are the applications of comparator?

- Zero crossing detector

- Window detector

- Time marker generator

- Phase detector

5. What is a Schmitt trigger?

Schmitt trigger is a regenerative comparator. It converts sinusoidal input into a square wave output. The output of Schmitt trigger swings between upper and lower threshold voltages, which are the reference voltages of the input waveform.

6. What is a multivibrator?

Multivibrators are a group of regenerative circuits that are used extensively in timing applications. It is a wave shaping circuit which gives symmetric or asymmetric square output. It has two states either stable or quasi- stable depending on the type of multivibrator.

7. What do you mean by monostable multivibrator?

Monostable multivibrator is one which generates a single pulse of specified duration in response to each external trigger signal. It has only one stable state. Application of a trigger causes a change to the quasi-

stable state. An external trigger signal generated due to charging and discharging of the capacitor produces the transition to the original stable state.

8. What is an astable multivibrator?

Astable multivibrator is a free running oscillator having two quasi-stable states. Thus, there are oscillations between these two states and no external signal is required to produce the change in state.

9. What is a bistable multivibrator?

Bistable multivibrator is one that maintains a given output voltage level unless an external trigger is applied. Application of an external trigger signal causes a change of state, and this output level is maintained indefinitely until an second trigger is applied . Thus, it requires two external triggers before it returns to its initial state.

10. What are the requirements for producing sustained oscillations in feedback circuits?

For sustained oscillations,

‾ The total phase shift around the loop must be zero at the desired frequency of oscillation, fo. ie, LAB =0 (or) 360º

‾ At fo, the magnitude of the loop gain | A b | should be equal to unity

11. What are the different types of filters?

Based on functions: Low pass filter, High pass filter, Band pass filter, Band reject filter

Based on order of transfer function: first , second, third higher order filters.

Based on configuration: Bessel, Chebychev, Butterworth filters.

12. List the broad classification of ADCs.

1. Direct type ADC.

2. Integrating type ADC.

13. List out the direct type ADCs.

1. Flash (comparator) type converter

2. Counter type converter

3. Tracking or servo converter

4. Successive approximation type converter

14. List out some integrating type converters.

1. Charge balancing ADC

2. Dual slope ADC

15. What is integrating type converter?

An ADC converter that perform conversion in an indirect manner by first changing the analog I/P signal to a linear function of time or frequency and then to a digital code is known as integrating type A/D converter.

16. Explain in brief the principle of operation of successive Approximation ADC.

The circuit of successive approximation ADC consists of a successive approximation register (SAR), to find the required value of each bit by trial & error. With the arrival of START command, SAR sets the MSB bit to 1. The O/P is converted into an analog signal & it is compared with I/P signal. This O/P is low or High. This process continues until all bits are checked.

17. What are the main advantages of integrating type ADCs?

i. The integrating type of ADC's do not need a sample/Hold circuit at the input.

ii. It is possible to transmit frequency even in noisy environment or in an isolated form.

18. Define conversion time.

It is defined as the total time required to convert an analog signal into its digital output. It depends on the conversion technique used & the propagation delay of circuit components. The conversion time of a successive approximation type ADC is given by

$$T(n+1)$$

where T----clock period

Tc---conversion time   n----no. of bits

19. Define resolution of a data converter.

The resolution of a converter is the smallest change in voltage which may be produced at the output or input of the converter. Resolution (in volts)= VFS/2n-1=1 LSB increment. The resolution of an ADC is defined as the smallest change in analog input for a one-bit change at the output.

20. Explain in brief stability of a converter

The performance of converter changes with temperature age & power supply variation. So all the relevant parameters such as offset, gain, linearity error & monotonicity must be specified over the full temperature & power supply ranges to have better stability performances.

21. What is meant by linearity?

The linearity of an ADC/DAC is an important measure of its accuracy & tells us how close the converter output is to its ideal transfer characteristics. The linearity error is usually expressed as a fraction of LSB increment or percentage of full-scale voltage. A good converter exhibits a linearity error of less than ±½ LSB.

22. What is a sample and hold circuit? Where it is used?

A sample and hold circuit is one which samples an input signal and holds on to its last sampled value until the input is sampled again. This circuit is mainly used in digital interfacing, analog to digital systems, and pulse code modulation systems.

23. Define sample period and hold period.

The time during which the voltage across the capacitor in sample and hold circuit is equal to the input voltage is called sample period. The time period during which the voltage across the capacitor is held constant is called hold period.

Special ICs

1. What are the applications of 555 Timer?

· astable multivibrator

· monostable multivibrator

· Missing pulse detector

· Linear ramp generator

· Frequency divider

· Pulse width modulation

· FSK generator

· Pulse position modulator

· Schmitt trigger

2. List the applications of 555 timer in monostable mode of operation

- missing pulse detector
- Linear ramp generator
- Frequency divider
- Pulse width modulation.

3. List the applications of 555 timer in Astable mode of operation

* FSK generator

* Pulse-position modulator

4. Define 555 IC?

The 555 timer is an integrated circuit specifically designed to perform signal generation and timing functions.

5. List the basic blocks of IC 555 timer?

· A relaxation oscillator

· RS flip flop

· Two comparator

· Discharge transistor.

6. List the features of 555 Timer?

· It has two basic operating modes: monostable and astble

· It is available in three packages. 8 pin metal can , 8 pin dip, 14 pin dip.

· It has very high temperature stability.

7. Define duty cycle?

The ratio of high output and low output period is given by a mathematical parameter called duty cycle. It is defined as the ratio of ON Time to total time.

8. Define VCO.

A voltage controlled oscillator is an oscillator circuit in which the frequency of oscillations can be controlled by an externally applied voltage.

9. List the features of 566 VCO.

· Wide supply voltage range (10-24V)

· Very linear modulation characteristics

· High temperature stability

10. What is meant by PLL?

A PLL is a basically a closed loop system designed to lock output frequency and phase to the frequency and phase of an input signal.

11. Define lock range.

When PLL is in lock, it can trap freq changes in the incoming signal. The range of frequencies over which the PLL can maintain lock with the incoming signal is called as lock range.

12. Define capture range.

The range of frequencies over which the PLL can acquire lock with the input signal is called as capture range.

13. Define pull-in time.

The total time taken by the PLL to establish lock is called pull-in time.

14. List the applications of 565 PLL.

· Frequency multiplier

· Frequency synthesizer

· FM detector

Application ICs

1. What is a voltage regulator?

A voltage regulator is an electronic circuit that provides a stable dc voltage independent of the load current, temperature, and ac line voltage variations.

2. Give the classification of voltage regulators:

- Series / Linear regulators
- Switching regulators.

3. What is a linear voltage regulator?

Series or linear regulator uses a power transistor connected in series between the unregulated dc input and the load and it conducts in the linear region .The output voltage is controlled by the continuous voltage drop taking place across the series pass transistor.

4. What is a switching regulator?

Switching regulators are those which operate the power transistor as a high frequency on/off switch, so that the power transistor does not conduct current continuously. This gives improved efficiency over series regulators.

5. What are the advantages of IC voltage regulators?

- low cost
- high reliability
- reduction in size
- excellent performance

6. Give some examples of monolithic IC voltage regulators:

78XX series fixed output, positive voltage regulators

79XX series fixed output, negative voltage regulators

723 general purpose regulator.

7. What is the purpose of having input and output capacitors in three terminal IC regulators?

A capacitor connected between the input terminal and ground cancels the inductive effects due to long distribution leads. The output capacitor improves the transient response.

8. Define line regulation.

Line regulation is defined as the percentage change in the output voltage for a change in the input voltage. It is expressed in millivolt or as a percentage of the output voltage.

9. Define load regulation.

Load regulation is defined as the change in output voltage for a change in load current. It is expressed in millivolt or as a percentage of the output voltage.

10. What is meant by current limiting?

Current limiting refers to the ability of a regulator to prevent the load current from increasing above a preset value.

11. Give the drawbacks of linear regulators:

- The input step down transformer is bulky and expensive because of low line frequency.
- Because of low line frequency, large values of filter capacitors are required to decrease the ripple.
- Efficiency is reduced due to the continuous power dissipation by the transistor as it operates in the linear region.

12. What is the advantage of switching regulators?

- Greater efficiency is achieved as the power transistor is made to operate as low impedance switch.
- Power transmitted across the transistor is in discrete pulses rather than as a steady current flow.
- By using suitable switching loss reduction technique, the switching frequency can be increased so as to reduce the size and weight of the inductors and capacitors.

13. What is an opto-coupler IC?

Opto-coupler IC is a combined package of a photo-emitting device and a photo sensing device

Contents

www.ingramcontent.com/pod-product-compliance
Lightning Source LLC
Chambersburg PA
CBHW061521180526
45171CB00001B/276